Gaines, Ann Graham.
Satellite communication 000798

384/.51/GAI

RSN=91034911   LOC=aub

SATELLITE COMMUNICATIONS

Published by Smart Apple Media
123 South Broad Street
Mankato, Minnesota 56001

Copyright © 2000 Smart Apple Media.
International copyrights reserved in all countries. No part of this book may be reproduced in any form without written permission from the publisher.
Printed in the United States of America.

Photos: page 7–CORBIS/Bettmann; page 12–CORBIS; pages 15, 18–CORBIS/Bettmann

Design and Production: EvansDay Design

Library of Congress Cataloging-in-Publication Data
Gaines, Ann Graham.
Satellite Communications / by Ann Graham Gaines
p. cm. – (Making contact)
Includes index.
Summary: Examines the development of telecommunication satellites and discusses their many uses.
ISBN 1-887068-63-5
1. Artificial satellites in telecommunication—Juvenile literature. [1. Artificial satellites in telecommunication.] I. Title. II. Series: Making contact (Mankato, Minn.)

TK5104.G35   1999
384.5'1—dc21   98-20889

First edition

9 8 7 6 5 4 3 2 1

SATELLITE COMMUNICATIONS

MAKING CONTACT

ANN GRAHAM GAINES

OBSERVERS OF THE NIGHT SKY sometimes notice bright lights that don't appear on any star chart. Often, these lights are **artificial satellites**. Many of these orbiting objects are **communications satellites**, which connect long-distance telephone calls across the world or relay television signals. Using **radio-frequency signals**, these satellites also relay faxes, telex messages, and digital **data**. Satellite communications is one of space technology's greatest achievements and its biggest commercial, or moneymaking, business. Developed over the last 60 years, this marvel of space technology now benefits millions of people around the world every day.

Launching an Idea

Astronomers first used the word "satellite" to describe a celestial body—moon, star, or planet—that orbits another. Earth and its moon are both satellites, since the earth orbits the sun and the moon orbits the earth. Today, hundreds of artificial satellites made and launched by humans are also orbiting the earth.

People have observed the moon with fascination for thousands of years. Although many ancient civilizations studied the different phases of the moon, they had no way of knowing that the moon orbits the earth. Copernicus, a Polish astronomer in the early 1500s, was the first to understand the relationship between the moon and the earth.

But Copernicus thought that the sun and all the planets in our solar system orbited the earth—that Earth was the center of our universe. It was not until the late 17th century that mathematician Sir Isaac Newton offered an accurate description of the way in which celestial bodies travel. According to Newton's law of gravity, the sun pulls on the earth, and the earth pulls back on the sun. It is only their constant sideways motion that keeps celestial bodies from crashing into one another.

> In 1687, English mathematician Sir Isaac Newton's law of gravity led him to believe that placing an artificial satellite into orbit was possible. However, the rocket technology needed to carry a satellite into space wouldn't be available for another 270 years.

ARTIFICIAL SATELLITE

any man-made spacecraft launched into orbit around the earth

COMMUNICATIONS SATELLITE

a satellite built to receive and transmit messages or information

RADIO-FREQUENCY SIGNALS

signals carried through the earth's atmosphere by electromagnetic waves

DATA

information; it can be sent worldwide using satellites

✳ COPERNICUS, THE FIRST ASTRONOMER TO DISCOVER THAT THE MOON ORBITS THE EARTH.

Soon after Newton published his law of gravity, people began exploring the possibility of launching an artificial satellite into space. It was an intriguing idea, but the technology needed to launch a satellite was still far beyond reach. In order to avoid falling back to the earth, a satellite has to travel at a minimum speed of 17,000 miles (27,370 km) per hour. For centuries after Newton's time, such velocity was impossible to achieve.

In the 1940s, writer Authur C. Clarke published an article in a magazine called *Wireless World*. In the article, Clarke explained that a satellite orbiting at an **altitude** of 22,300 miles (35,900 km) above the equator would circle the earth in exactly 24

In 1955, the Soviet Union and the United States announced plans to launch small, experimental satellites. The successful orbits of *Sputnik 1* and *Score* generated intense optimism in the field of satellite communications.

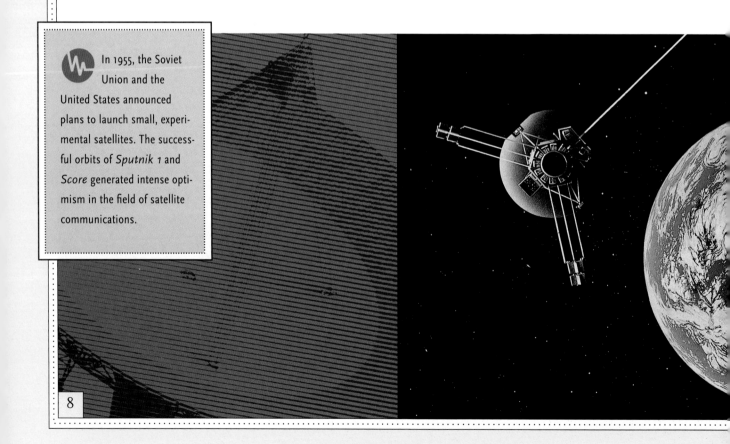

hours. Such a satellite would maintain a **geostationary orbit.** To observers on Earth, it would appear to be staying motionless in the sky. Clarke also proposed that people might one day launch three manned satellites into a geostationary orbit. By placing three satellites into space, scientists would be able to form a giant triangle of communication by which people on Earth could transmit television programs and telephone signals around the world.

As is the case with many revolutionary ideas, Clarke's article attracted little attention when it was published. However, German scientists took the idea to heart and began investigating the possibility of using satellites as weapons during World

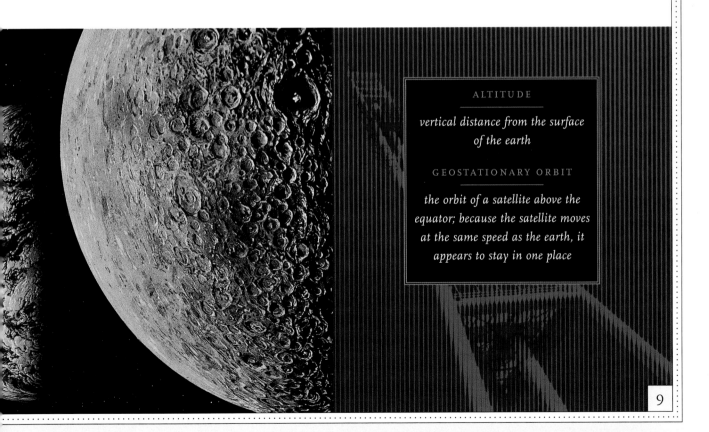

ALTITUDE

vertical distance from the surface of the earth

GEOSTATIONARY ORBIT

the orbit of a satellite above the equator; because the satellite moves at the same speed as the earth, it appears to stay in one place

✶ LIKE THE MOON, MAN-MADE SATELLITES ORBIT EARTH, WHICH IN TURN ORBITS THE SUN.

War II. Although its goal was military supremacy, Germany's research led to the technology needed to launch an artificial satellite, an achievement that would ultimately benefit the world.

In 1955, John Pierce, who worked in the laboratories of the huge AT&T telephone company, wrote an article explaining how satellites might eventually be used for communication. Pierce was especially interested in learning whether companies could make money by providing satellite communications.

The Soviet Union launched *Sputnik 1*, the world's first artificial satellite, in October 1957. *Sputnik 1* was a small sphere that weighed only 184 pounds (83 kg) and measured 22 inches (56 cm) across. It orbited just a few hundred miles above the earth. At such a low altitude, *Sputnik 1* could complete an entire trip around the planet in just 90 minutes. The satellite carried a radio transmitter and broadcast a steady beeping sound that allowed Soviet scientists to monitor its course. Although *Sputnik 1* transmitted radio signals, it cannot be considered a true communications satellite since it was not used to relay a message between two places.

The Soviets launched *Sputnik 1* mainly to demonstrate their technological progress and to establish supremacy in space. When Americans first heard the news of its successful launch,

> The Soviet Union launched the first satellite of its *Molniya* communications program on April 23, 1965. Two more series of *Molniya* satellites followed within the next 10 years. *Molniya 1S*, launched as part of the third series on July 29, 1974, was the first Soviet geostationary satellite in space.

many were frightened. If the Soviet Union was capable of orbiting satellites above the United States, what was to stop it from bombing the U.S. with satellite-dropped missiles? Although concerned Americans didn't know it at the time, their fears were largely ungrounded—even with today's technology, it is very difficult to accurately drop a bomb from a satellite. However, Americans now desperately wanted to catch up with the Soviet Union in the field of space technology. The United States launched its first artificial satellite, *Explorer 1*, in 1958.

By this time, both the Americans and the Soviets were busy trying to develop and launch communications satellites. For years, underwater cables laid across the ocean floor had allowed

* *SPUTNIK 1*—A SMALL, LOW-FLYING OBJECT—BECAME THE FIRST ARTIFICIAL SATELLITE IN SPACE IN 1957.

people to send telegraph messages between Europe and the United States. In 1956, workers finished building the first transatlantic telephone cable, allowing Americans and Europeans to place telephone calls to one another. But these cable communications could be unreliable, and the transatlantic cable proved to have numerous flaws almost immediately, the main one being that it was not able to carry many calls.

People could already send radio signals over very long distances, because these signals bounced off the **ionosphere** and kept going close to the earth's surface. But the same couldn't be said for the new communications medium called television. Instead of bouncing off the ionosphere, TV sig-

* UNDERWATER TELEPHONE CABLES HAVE CONNECTED EUROPE AND NORTH AMERICA SINCE 1956.

nals stayed close to the earth for about 150 miles (250 km) and then disappeared into space. In order to cover international events during the 1950s, television networks had to fly motion picture film of the events from one country to another before any pictures could be broadcast. It was a painfully slow process that delayed coverage of important breaking news.

For these reasons and many others, people living in the late 1950s wanted a better long-distance communications system. Artificial satellites promised to be the answer to their problems.

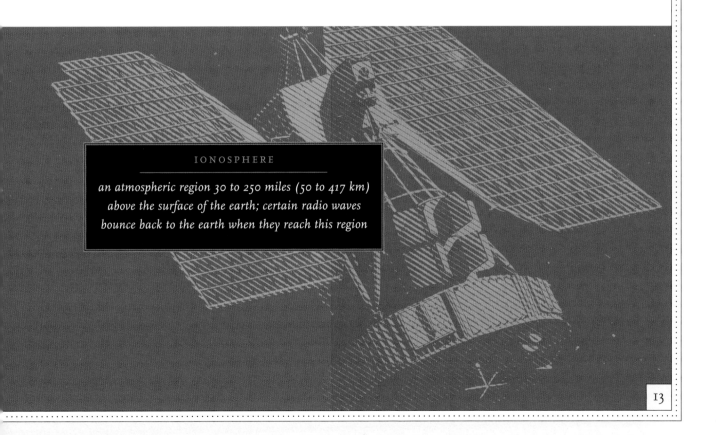

IONOSPHERE

an atmospheric region 30 to 250 miles (50 to 417 km) above the surface of the earth; certain radio waves bounce back to the earth when they reach this region

Global Communications

The United States soon stepped forward as a leader in developing satellite communications. In the 1950s, scientists at both **NASA** and the U.S. Department of Defense worked on communications satellites. In 1958, the U.S. launched *Score*, a satellite carrying a tape-recorded Christmas message from President Dwight D. Eisenhower. The message was broadcast from the satellite as it passed over designated **groundstations**. Although historians sometimes refer to *Score* as the first communications satellite launched into space, it did not fully function as one since it could not receive or relay messages. The usefulness of the satellite was limited as well by its low orbit, which made *Score* visible to groundstations for only a few minutes at a time.

In 1960, NASA launched a simple, experimental **passive communications satellite.** *Echo 1* was simply a huge plastic balloon coated with aluminum. It measured 100 feet (30.5 m) in diameter and maintained a low orbit 930 miles (1,500 km) above the earth. *Echo 1* made it possible to communicate long distances by reflecting telephone and television signals that were beamed at it. However, scientists soon decided that passive systems would not

> The first direct broadcasting satellite, *ATS 6*, was launched on May 30, 1974. This satellite laid the groundwork for satellite-assisted television broadcasts.

NASA

the National Aeronautics and Space Administration; a U.S. government agency that develops and tests space technology

GROUNDSTATION

a location on Earth that uses communications satellites as relay points to send and receive information

PASSIVE COMMUNICATIONS SATELLITE

a satellite that reflects signals back to Earth without amplifying them

✴ *ECHO 1*, A GIANT SPHERICAL SATELLITE DESIGNED TO REFLECT POWERFUL RADIO SIGNALS BACK TO EARTH.

work. The systems required very powerful **transmitters** and large ground **antennas**, which made them extremely expensive. In addition, both *Echo 1* and *Score* presented the same problem: they circled the earth at a very low altitude, so groundstation managers had a very short time in which to receive transmissions from the satellites.

A satellite called *Courier 1B* also entered space in 1960. This was an **active communications satellite** that could actually receive and transmit messages instead of just reflecting them. It also had severe limitations, since it could relay only teletype messages and could not transmit television programs or telephone calls.

In 1962, NASA launched *Telstar* for the AT&T company. This satellite's

* TOWERING ANTENNAS SUCH AS THIS ONE AT A TELEVISION STATION SEND AND RECEIVE BROADCAST SIGNALS.

medium-range orbit was elliptical, passing within 620 miles (1,000 km) of the earth at one end but 3,720 miles (6,000 km) from the planet at the other end. At one point along its orbit, people in both Europe and the United States could see *Telstar*. At another point, people in the United States and Japan could spot the satellite at the same time. As a result, *Telstar* provided **real time** communication between the United States, Europe, and Japan, if only for a few minutes at a time.

ACTIVE COMMUNICATIONS SATELLITE

a satellite that amplifies the signals it receives, converting the messages to different frequencies before retransmitting them

REAL TIME

the actual time during which something takes place

TRANSMITTER

an electronic device that sends signals to a receiving site

ANTENNA

a device used to receive or transmit radio waves

✻ MODERN SATELLITES, EQUIPPED WITH POWERFUL TRANSMITTERS, CAN RELAY INFORMATION ACROSS THE WORLD IN SECONDS.

The first communications satellite launched into a geostationary orbit was *Syncom 2*. NASA launched and operated this satellite for the U.S. government. Orbiting over the equator at an altitude of 22,300 miles (35,900 km), *Syncom 2* could be seen by almost half the people on Earth at the same time. Thanks to this satellite and others like it, Americans were able to watch live coverage of the 1964 Olympics in Japan.

On April 6, 1965, an American company called the Communications Satellite Corporation, or COMSAT, launched its first satellite. *Early Bird,* which was later renamed *Intelsat I,* was the first commercial communications satellite. When *Early Bird* began

* *Early Bird (Intelsat I),* the first communications satellite launched by a business.

communicating with other satellites in geostationary orbits, the goal of global communication was almost within reach.

In 1969, satellites in the *Intelsat III* series provided communication for the first time to groundstations around the Indian Ocean. This link completed global coverage by satellite communications. On July 20, 1969, just a few days after this final global link, 500 million people around the world watched *Apollo 11* land on the moon. Communications satellites relayed the historic pictures across thousands of miles of space and into the homes of everyone with a television.

An orbiting satellite was repaired for the first time in April 1984. After a *Challenger* space shuttle astronaut stopped the satellite from spinning, a robotic arm from the shuttle lifted the satellite into the cargo bay, where astronauts replaced a defective fuse.

* TELEVISION BROADCASTS OF THE WORLD'S FIRST MOON LANDING WERE MADE POSSIBLE BY SATELLITES.

The Growth of an Industry

The Communications Satellite Corporation, founded in 1963, was the first company to launch and operate communications satellites on a commercial basis. A non-profit organization called the International Telecommunications Satellite Organization, or INTELSAT, was formed the following year and soon became the world's largest satellite communications system. By the late 1990s, the organization provided service to about 200 nations, and about 140 countries had invested in INTELSAT.

By the early 1990s, six companies were providing round-the-clock satellite communications to the United States through 36 operating satellites. A system called Pan American Satellite, or PanAmSat, had emerged to compete with INTELSAT by offering international service. The satellite communications industry also began to include more satellite manufacturing companies and service providers.

Communications satellites that are not owned by private companies are generally owned by national governments. Many countries began developing their own communications satellites during the 1970s. The Soviet Union, Canada, and the United States were the

> An orbiting satellite was brought back to Earth for the first time in November 1984. Astronauts used the *Challenger* space shuttle's robotic arm to retrieve two communications satellites from bad orbits.

✳ Satellites placed into high orbits are often carried from Earth by powerful rockets.

first nations to develop satellite systems, but other countries have also used the growing technology to build satellites of their own.

The country of Indonesia was the fourth nation to launch a publicly owned communications satellite. This nation is made up of 13,677 islands covering 735,555 square miles (1,904,570 sq km) in the Pacific Ocean. Because its citizens are so widely scattered across the islands, it was critical that Indonesia develop a wider-reaching means of communication.

Some governments are responsible for the transmission of television programs and telephone calls to their citizens. Many government agencies depend on satellites

to relay great quantities of digital data between computers. Military forces also depend on satellite communications to send messages from place to place. To prevent enemies from intercepting important information, many of these messages are sent in secret codes.

The International Maritime Satellite Organization, or INMARSAT, also operates a number of communications satellites. INMARSAT has built a mobile satellite telecommunications network to provide better communications service to ships at sea and offshore facilities such as oil rigs.

By 1995, governments, private companies, and other specialized organizations had launched about 4,500 satellites of all types. Roughly 2,200 remained in orbit, although some of

✳ SATELLITES MAKE RELIABLE COMMUNICATION POSSIBLE BETWEEN ISLANDS SUCH AS THOSE OF INDONESIA.

them were no longer working properly. In the late 1990s, there were about 200 functioning communications satellites in orbit. Ten to 20 additional communications satellites go into orbit each year, and two-thirds of them are placed in a geostationary orbit. Between 1965 and 1997, INTELSAT alone launched close to 60 communications satellites. Those in the latest series can transmit 22,000 telephone calls and three color television broadcasts simultaneously.

Many companies have made huge profits by operating satellites. Governments, telephone companies, and television networks are willing to pay enormous amounts of money to use the amazing capabilities of communications satellites. Because the making and launching of satellites is an extremely expensive process, satellite-making companies have to charge their customers high rates. A single large communications satellite costs about 75 million U.S. dollars, and the launching devices that send a satellite from Earth into orbit cost about the same amount. Making and launching satellites can also be a risky business. After a lot of time and money have been invested in their construction and launch, some satellites never achieve their planned orbits, and others break down or quit working prematurely.

> The *Intelsat VI* series, begun in 1989, represented a big step forward in the capabilities of satellite communications. Each satellite, which weighed about five and a half tons (5 t), could relay between 33,000 and 100,000 telephone conversations at a time.

* Launching multimillion-dollar satellites into space is an expensive process.

How Communications Satellites Work

In the early days of satellite technology, guided missiles carried satellites into orbit. In 1998, satellites were carried by either multi-stage rockets or by space shuttles, which place the satellites in the correct position and altitude and then let gravity take them into orbit.

The shell of a communications satellite is called its bus. Balloon-shaped buses were tried in 1960, but today most buses are shaped like cylinders or boxes. The communications equipment carried by a bus is called the payload.

Communications satellites need a source of electrical power to operate this equipment. To generate the necessary power, the satellites use large solar panels that resemble wings. These panels are covered with solar cells that absorb sunlight and convert it into electricity. Satellites also carry batteries to back up their main solar power source.

An uplink and a downlink are basic parts of every communications satellite. The uplink includes a receiving antenna and a receiver, which allow the satellite to receive messages from groundstations on Earth. The downlink consists of a transmitting antenna and a transmitter, which allow the satellite to send mes-

> The major Earth groundstations that communicate with satellites have a receiving antenna about 100 feet (30 m) in diameter, as well as a powerful radio transmitter to send signals out into space.

26

sages back to one or more groundstations. In some cases, a single antenna can both receive and transmit messages. Inside the satellite are electronic circuits that store data. **Amplifiers** attached to transmitters strengthen or change the frequency of a received message before the satellite relays the message back down to Earth.

Satellites do not actually relay sound. Instead, telephones convert sound to electrical signals that are sent up to satellites as radio waves. These waves are then sent back to Earth, where they are translated back into sound by telephones on the receiving end.

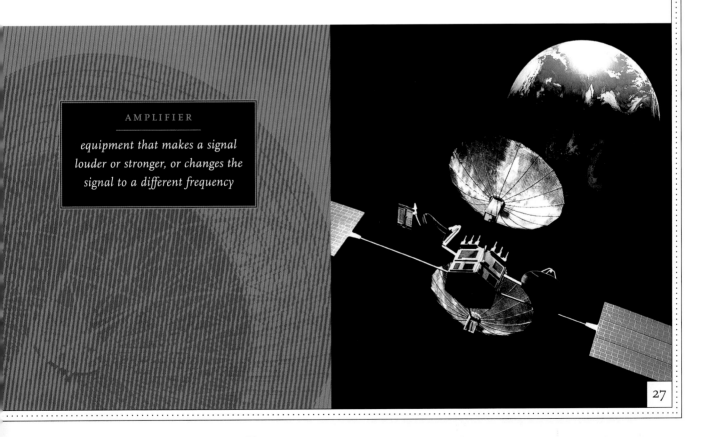

AMPLIFIER

equipment that makes a signal louder or stronger, or changes the signal to a different frequency

✳ TODAY, HUNDREDS OF SATELLITES ARE IN CONSTANT ORBIT AROUND THE EARTH.

Early satellite developers worried that telephone callers would be annoyed by the length of time needed to transmit a call from a groundstation to a satellite and back to another groundstation. But most callers don't even notice the slight delay as the satellite relays the telephone signal.

Up until the mid-1990s, telecommunications satellites relayed TV signals only to large, expensive **satellite dishes** on Earth. Since 1994, however, direct broadcasting satellites have transmitted TV programs to smaller, less costly dishes measuring just 18 inches (46 cm) in diameter.

> In 1995, Pan American Satellite entered the satellite communications industry. It became the first private company to provide global satellite communications services.

※ WIRELESS TELEPHONE COMMUNICATION IS ONE OF THE MOST WIDESPREAD BENEFITS OF SATELLITES.

SATELLITE DISH

a dish-shaped antenna that picks up television signals from communications satellites

✶ DISH-SHAPED ANTENNAS HELP SEND AND RECEIVE RADIO WAVES OVER THOUSANDS OF MILES.

The Future of Satellite Communications

Today, satellite designers are testing several new approaches to improving satellite communications. NASA has launched an Advanced Communications Technology Satellite, or ACTS, which carries many uplinks and downlinks. Each link covers an area only a few hundred miles wide, but it can be aimed quickly and accurately.

Another proposed approach would involve launching a "constellation" of low-orbiting satellites that would cover the earth. This approach is hampered by the fact that satellites are so expensive to build and launch. Yet some companies are looking into the mass production of communications satellites. This would reduce the construction cost of each satellite, making this idea a possibility.

In the future, communications satellites will continue to provide convenient, worldwide service in a number of ways. Satellite relay of television signals will make a wide array of programs available to more and more people. The use of satellites with mobile communications devices, such as pagers and cellular telephones,

> In the late 1990s, the United States used two main spacecraft launch sites: the Kennedy Space Center in Florida, and Vandenberg Air Force Base in California. Most geostationary communications satellites are launched from Florida, which is closer to the equator.

is becoming more widespread. Governments and private companies will use satellites to transfer greater amounts of data across long distances. Internet users will be connected via satellite, and communications satellites may soon help travelers in all parts of the world reach their destinations much easier by providing quick, detailed routes.

In the coming decades, satellite communications will play a bigger and more important role in our everyday lives. The global flow of information, conversation, and entertainment is an ever-increasing one in a world whose population continues to grow. Much of that flow is routed through those bright, orbiting lights in the night sky.

✳ CELLULAR PHONES AND WIRELESS COMPUTER CONNECTIONS REPRESENT
THE FUTURE OF GLOBAL COMMUNICATION.

INDEX

A
Advanced Communications Technology Satellite (ACTS) 30
Apollo 11 19
AT&T 10, 16
ATS 6 14

C
Canada 20
Challenger 19, 20
Clarke, Authur C. 8–9
Communications Satellite Corporation (COMSAT) 18–19, 20
communications satellites 5, 7, 9–10, 11, 14, 16–19, 20–24, 26–28, 30–31
  active 16, 17
  broadcasting 14
  design 26–27
  launching of 26
  passive 14–16
Copernicus 6, 7
Courier 1B 16

E
Early Bird 18
Echo 1 14, 15, 16
Eisenhower, Dwight D. 14
Europe 12, 17
Explorer 1 11

G
geostationary orbit 9
Germany 9–10

gravity 6, 8
groundstations 14, 15, 26–27, 28

I
Indonesia 22–23
Intelsat III 19
Intelsat VI 24
International Maritime Satellite Organization (INMARSAT) 23
International Telecommunications Satellite Organization (INTELSAT) 20, 24

J
Japan 17

K
Kennedy Space Center 30

M
mobile communications devices 30–31
Molniya 10
Molniya 1S 10

N
NASA 14, 15, 16, 18, 30
Newton, Isaac 6–8

P
Pan American Satellite (PanAmSat) 20, 28
Pierce, John 10

R
radio-frequency signals 5, 7, 12

S
satellite dishes 28, 29
Score 8, 14, 16
Soviet Union 8, 10–11, 20–22
Sputnik 1 8, 10, 11
Syncom 2 18

T
telegraph 12
telephones 22, 24, 27–28
television 19, 24
television signals 12–13, 28
Telstar 16–17
transatlantic cables 12

U
U.S. Department of Defense 14

V
Vandenberg Air Force Base 30

W
Wireless World 8
World War II 9–10

AUBREY SCHOOL
1075 Stratford Avenue
Burnaby, B. C.
V5B 3X9